The Essential IT Contractor's Survival Guide

by Courtney Thompson

The Essential IT Contractor's Survival Guide
©2003 Courtney Thompson

ISBN: 1-4116-0489-X

Published by:
Courtney Thompson
Courtneythompson2003@yahoo.com

This & other books by this author are available your local
bookstore, or direct from:

www.lulu.com/courtneythompson

Special thanx (don't you just love literary
license?!) to Sheila Lawrence, my editor & Debbie
Dunford, a Senior Recruiting Manager with a
fortune 500 IT Staffing Services firm. I thank
them for their time, advice and friendship, to help
me create this book for you. I say again to these
two dedicated ladies - *thanx*.

Printed in the United States of America

Table of Contents

1.0   READ ME FIRST!!!

Reading this chapter first will help you get the most out of this book. This text was written with busy people like you in mind. This book is designed to be brief and informative.

This book was written for computer professionals and recent graduates who want or need to work on a contract basis. The material presented is designed to help those new to computer contracting to start out right; as well as aid those with years of experience to improve aspects of how they approach contract work. Note that the use of terms: computer contractor or Information Technology (IT) contractor are used throughout this book and are treated the same. Freelancer is another commonly used term by some for the profession.

Working on contract is not new. Many types of employment such as carpenters, plumbers, repair-people, lawyers and architects are usually hired on a fee for service (or contract) basis. The contract usually has a specific start and end date. What is new is the fact that many who have been used to working a regular job are now faced with changing their work environment on a more frequent basis.

Audience for this Book:

This book is intended for anyone who wants to pursue a career as an IT contractor. The information contained within these pages is designed to aid:

- New entrants to the IT workforce
- Women returning to work
- Those seeking a career change
- Professionals with disabilities
- Victims of corporate downsizing

Getting the Most from this Book:

Questions are presented at the conclusion of each chapter. They are designed to get you thinking. They will push and challenge what you currently know. Your present thoughts are what are sought after here: rather than what can be termed as a correct answer.

Chapter Review:

1-1) In a short paragraph, describe what contract work is.

1-2) Do you want to work as a computer contractor? Why?

1-3) What things must a work assignment have to keep you interested?

1-4) Do you enjoy learning new things? Discuss or write about 2 of them.

1-5) From what you know today; are you afraid of becoming an IT contractor? Why?

1-6) Why do you want to work in information technology (IT)?

Chapter Notes

2.0   The Nature of Contract IT Work

Working on contract means working on a fee for service basis. Although this concept seems new; it is actually a return to the days of old. In the days before the industrial revolution, most people (millers, blacksmith, weavers, carpenters, knights) worked on a fee for service or contract basis. The more skilled one was at his or her chosen craft, the better their monetary compensation, due to supply verse demand. They were also known for their good character (i.e.- Joshua the mason was always honest with you), the financial rewards (or at least frequency of steady work) are there too. *[Refer to ch. 7 on 'Professionalism']*

Contract work in the Information Technology (IT) world is very different from most other fields of endeavor; be it a trade (i.e.- carpentry, plumbing) or a profession (i.e.- engineer, lawyer). This is due to the fact that technology keeps on changing at a furious pace. Computing power doubles approximately every 18 months. With the rapid changes, come just as rapid demands for IT professionals to constantly update ones skills while still performing their daily tasks.

An important question arises. Why do many firms use IT contractors as opposed to hiring permanent staff? The short answer is flexibility. The benefits to a firm's IT manager are many. Examining them in detail will help you better understand the nature of contract work and provide you with some insight as to how to utilize contract skills if and when you may be in a position to do so. The benefits to the IT manager include:

1) Staff size flexibility;
2) Reduction of training costs;
3) Reduced salary overhead.

Here is one scenario. An IT manager is directed, by senior management, to change the network to a new operating system, within 90 days. Oh! We forgot the caveat, the network has 2,200 users at 5 locations in three cities; there is a large budget to do whatever needs to be done. What would you do in this situation? Many would hire more IT staff. The problem with that is the 20 people you hired for the migration project will have nothing to do once the conversion is over. Day to day operations require far less staff (if the IT department is on the ball).

In this instance many IT managers will turn to IT contractors, to help accomplish this and other major projects. This is due to the flexibility a firm gains in the sizing and allotment of human resources to IT projects. As a project expands, a manager can increase his staff quickly by utilizing contract IT staff. As a short-term project ends, the manager can downsize just as quickly with the same ease.

The next point is that training costs (in most cases) is eliminated. This is because most contractors are hired for their expertise such as Peoplesoft, Oracle, NT, etc., are expected to be fully qualified for that task already. Therefore, no training is required. In effect, the contractor is brought in as a hired gun. Go in, fix the problem, create a solution, or execute a client devised plan. When the task is complete, our masked avenger rides off into the sunset on to the next task. This is the ideal situation. In reality sometimes training may be required *[Refer to ch.7 on 'Directing Your Career']*

Employees ask......I have a good job and I have been with my company for years, why do I need to know this stuff?

Well, times are a changing, as a good friend of mine puts it. The job for life that many baby-boomers had grown up to expect **does not exist** anymore. Corporate mergers are a way of life in the new millennium. It is conceivable that you could be the victim of a merger downsizing action at some point throughout your career. If you enjoy being an employee (which is fine), the goal is still, to have you start thinking like a contractor. By developing the skills contained herein will both prepare you for any future employment upheavals, and at the very least sharpen you mental attitude towards your current position.

Here are a few more ways full-time employees could use contracting. One can use agencies and contracting to make extra money (for holidays, vacations, kids' school, a new boat, more time to golf, etc.). One can also use contracting as a means to prepare for potential downsizing. When the rumors start flying getting yourself out there and connected with an experienced recruiter working for a reputable agency can really put you ahead of the other 40+ people in your department when the ax does swing. Those few days head start might be the difference between having a few days of unemployment versus several months.

Here are the other fringe benefits. Your resume will be properly updated; you will potentially have some new outside-your-fulltime-job-projects to put on your resume demonstrating your flexibility; you will have started a new relationship with a recruiter (or recruiters) that can help move you more quickly back into full-time work if you are laid off; you'll have a better idea about what other opportunities are floating around out there, etc. etc.

Chapter Review:

2-1)   How can the utilization of IT contractors help a firms dynamic staffing requirements?

2-2)   When might a firm consider using contract staff?

2-3)   When would a firm consider hiring a person for a permanent IT position?

2-4)   List the pros & cons of hiring a fulltime employee; discuss.

2-5)   List the pros & cons of hiring contract IT staff; discuss.

2-6)   How can contract work benefit you if you are a new IT graduate?

Chapter Notes

3.0   What are Your Skills Worth?

This is a loaded question. Everyone is skilled at something.
However, is anyone willing to pay you lots of cash for your
knowledge and skill? The question is: Are your skills
marketable? The answer to this question is: It depends.

Fact: *Seek to earn an hourly rate that averages 25-50%
more than what you earn for a similar fulltime
job.*

Remember that the nature of contract work is that it is
finite. It has a start and an end date. This also means that
after your current assignment ends, you may start another
one right afterwards, or you may be out of work for up to
several months. You need to plan and adjust your lifestyle
accordingly. *Refer to the chapter 6 'Money Matters' for
further discussion on managing your money.*

How do you go about finding out what a particular position
is worth? Use the internet to do a search for salaries. You
could try to search on some of the following:

- Company websites

- IT contracting sites; such as

    - www.computerjobs.com

    - www.monster.com

    - www.careerbuilder.com

    - www.hotjobs.com

    - www.dice.com

    - www.guru.com

- IT contracting agency website

    - You can find these via visiting some of the above websites

Maximizing your marketability is a balance of four key elements. They are:

1. Your possessed 'in demand' **Skill Set**
2. Your practical **Experience**
3. Your ongoing **Education**; coupled with
4. Strategic **Certification(s)**

Let's look at how each of these items plays a vital role in your career as an IT contractor.

Skill Set, specially an 'in demand' one is vital to your success. It is very important to get a firm handle on what you are interested in (and good at), that firms are willing to pay you top dollar for the privilege of utilizing your talents.

Experience is good. Experience in a vital area is best. That's because you were paid for your experience as long as the expertise you possess inset firm to improve its bottom line. Always seek ways to enhance your experience in your particular field of expertise. This especially for students, people changing careers or those returning to the workforce, understanding that simply getting a short term (1-6 month) assignment under their belt can be the difference between 'you' or the 'other guy' getting picked for the next contract.

Education is a constant in the IT world. You need to constantly train and enhance what you know in order to stay on top. There are several ways to keep your skills up to date without spending significant amounts of cash. In addition to taking courses, try some of the following ways of learning things:

1. Attend trade shows
2. Read related trade publications

3. To obtain books from the library on the subject

4. Surf the net for relevant information

Certification can be a potential boost to your career.
Depending on the field of information technology that you
specialize in, a certification tells people that you have a
solid grasp of the concept required to do with a particular
function. Check out the following websites to keep
informed about key certification(s):

- Certification Magazine www.certmag.com
- Computing Technology Industry Association
  (CompTIA) www.comptia.org
- Project Management Institute (PMI)
  www.pmi.org

Directing Your Career

One of the constants of being a computer contractor
(besides paying taxes) is change and lots of it. In some
professions, such as engineering, basic principals (i.e..-
Newtonian physics) haven't changed in decades (or
centuries). In IT what you knew about Java programming
and web browsers, for instance, two years ago, require a

little rethinking today; especially if you don't use this skill every day.

So, how do you keep up? First, you must decide what area you enjoy and want to specialize in. Some enjoy programming, network administration, network design, web development, project management, database management, disaster recovery design, or helpdesk. These are a few areas that are available. There are more areas of endeavor, many more. Since you cannot do them all and be successful at them, you must decide what you want to do.

How do you decide what area to focus on? Go back to basics. You really need to answer fundamental questions, not about technology, but about you. If you know what you want, like and can deal with, the rest just falls into place – really!

Chapter Review:

3-1) For each of the following IT position, find out the high, low and average salaries/wages for the following full-time (perm employee) positions: a) programmer; b) network administrator; c) desktop support analyst; d) project manager; e) network engineer; f) web developer; g) CAD operator; h) help desk analyst. Use the internet, interviews or articles to back up your answers.

3-2) Repeat the previous exercise and report how much those same specialists are worth on a contract basis.

3-3) For each of the positions listed in 3-1, is it worth doing that job on a contract basis? Why?

Chapter Notes

4.0   Tools of the Trade

As a professional you should be in possession of your own
set of tools. In the IT world, that translates to owning a
laptop computer capable of running the latest software.
Some may balk at the notion of forking out for their own
equipment. 'I tell you now, you believe me later' (thanks
Arnold). As with anything else in life you should be
chomping at the bit to ask the all-important question:

What's in it for Me?

First off, having your own equipment gives you a measure
of independence and flexibility. How so? You have a place
to play. When you want to test new ideas, recreate a
problem, play with new versions of software, you don't
have to seek (a.k.a. 'beg') permission to do so. Also, when
develop some really neat stuff (i.e.- some cool code), you
can add it to your bag of tricks.*[Refer to chapter 7
'Professionalism' for some point to watch for and to
ensure that you are doing the right thing.]*

Suggested Tools of our Trade

If you work anywhere in North America; here is a list of items that you **<u>must</u>** have:

1) Computer (preferably laptop or tablet PC)
2) Personal Digital Assistant (Palm or Pocket PC)
3) Cell phone
4) Personal email address

If you work with computers, you will need one, period. Would you hire a carpenter to work on your house if he didn't have his own tools and wanted to use whatever tools you had? That would not instill in you any confidence in his/her ability to do the work at hand. From years of experience (and travel), I strongly recommend getting a laptop or a tablet PC. Make sure that the unit you purchase has the capabilities to use the software packages that you need to use on a daily basis. These software tools may include (but are not limited to) the following:

1) An Office suite
2) Programming tools
3) Graphics tools

4) Network planning tools

Personal Digital Assistants (PDA) is a must for any IT professional. This is because it is a multifunction device that should do at the bare minimum:

1) Schedule tasks
2) Maintain address and phone numbers
3) Keep a To Do list
4) Create and edit short notes, passwords, network info, etc.

Note: For more info on how to utilize your PDA, check out the book entitled: '**Palm @ Work & Play**'; by Courtney Thompson; ISBN 1-4116-0198-X. It is available from: www.lulu.com/courtneythompson

Cell phone are not optional, they are required. Here is a justification list for having one:

1) Your contracting agency can reach you
2) You can call the client if an emergency arises
3) The client has a way of reaching you

4) Family can reach you in an emergency (never tie up your customer's phone)

With your own equipment you still need to obtain permissions in certain instances. For example: You can't tie-up your machine into a company's network without permission. You can't (shouldn't) copy their software to test on your machine without permission. You (especially if you are solo) might be opening yourself up to liabilities if you tie your machine into theirs and there is a virus or other problem. Customers may "always be right" but they are almost "always looking to save money" also – and sometimes that means finding a scapegoat. If you open the door, be sure you can pay to close it! Lastly, if you are working through an agency, using your own equipment does not get around the obligations that you agreed to when you signed on with your consulting company. Be sure you follow the computer policy agreements to the *letter* – especially if you work for multiple agencies. Make sure *this* job complies with *this* agency's policies.

Chapter Review:

4-1)   What advantages are there to owning your own equipment?

4-2)   Should you pay for your software? Explain your reasoning.

4-3)   Do some research and compare PDA models; 3 Palm OS bases and 3 Windows based.

4-4)   What factor(s) might determine which products you buy?

4-5)   Create a comparison table listing the pros and cons of: 1) Desktop PC; 2) Tablet PC; 3) Laptop

4-6)   Find out and write a short summary on any tax saving for purchasing your equipment and software.

Chapter Notes

5.0   Finding Contracts

There are two main ways to finding IT contracts: solo or via an agency. Naturally there are pros and cons to each. The method you choose will depend on the type of individual that you are; what tasks you enjoy doing; how contracts are obtained in a particular IT field or company.

The following sites are good starting points for those seeking jobs in Canada and the United States:

- www.computerjobs.com
- www.careerbuilder.com
- www.hotjobs.com
- www.dice.com
- www.monster.com
- www.guru.com

Obtaining Contract Assigns: Agency vs. Solo

	Agency	Solo
Pros	• Help with assignment locating • Help matching your skills to assignments • Use assignments for career change • Ensures that you are paid for services rendered • Possible benefits • 401K	• No middleman; negotiate your own terms • Entire fee is yours
Cons	• You get a portion of your billable rate • Being bill at a lesser rate make up for an agencies	• You have to handle promoting yourself • You have ensure that you get paid • You have to handle problems with the client • You may not be paid (i.e. client bankruptcy)

Dealing with Recruiters

You may work with technology, but it is people that you have to deal with to get, renew, modify or terminate a contract. Whatever firm you may go through to seek a new IT assignment, it is the individual recruiter that either makes or breaks you.

Keeping this in mind, you are not only seeking a placement firm, but also an individual whom your personality can blend and work with. I have a few recruiters that I have developed friendships with to the point that when the two of them changed firms, I moved with them. Why? It is much better to work with someone you know (and who knows you).

Think about it from this angle: If you have a position and have two equally qualified candidates, and one of the candidates is a friend of yours, who are you going to recommend for the position? If you are like most of us, you would favor your friend (it's human nature).

How a Recruiter Can Help

- *New Entrants to the IT Workforce*
  A good recruiter that specializes in placing entry level people can help one to create an effective resume and place them before a number of clients who are seeking such talent.

- *Women returning to work* This would be similar to the '*New Entrants to the IT Workforce*'. Using an agency can allow someone who only wants to return part-time do so without having to make the investment of being 100% on call (as would happen with contractors going solo and trying to get new customers). Also, women can utilize recruiters to see which elements of their previous life are most useful in obtaining a new position. They can help shape and guide the resume to highlight the 'right' pieces. She can work on a few short term assignments below her skill set just to get current experience on the resume (and eliminating an "old" date on the first line of the resume). If using an agency she would have a "backup" in place in case of problems at home – whereas a solo person re-entering the workforce is not so fortunate. Flexibility is also key, if she is unsure just how far 'back into the workplace' she wants to get. These elements are unique to someone who actually has experience but just hasn't used it in a while. Lastly, if a woman is returning to work she likely has at least some industry experience. A contracting company can sell the fact that this woman knows Banking or Credit Cards or Chemicals or Pharmaceuticals, etc. That might be just what she needs to get back in the door.

- *IT Contractors Seeking a Career Change*

  This is where your relationship built with a recruiter can work for you. Work with your recruiter to progressively move from one career path to another. I start in IT as a CAD technician. I let my recruiter know my desire to get into networking. He got me assignments on helpdesk, entry level desktop and then networking. On one contract, my recruiter got the client (IBM) to pay for network training. I could not have made this transition as smoothly without my recruiters help. He did that because we had developed a good professional relationship and friendship. Remember: *People are important.*

- *Professionals with disabilities*

  A recruiter can seek out good opportunities and 'good fits' for IT professionals with certain disabilities. Yeah, we know that the law says it is illegal to discriminate based on disability, race or sex. The reality is that discrimination is still alive and well (unfortunately) in most places in Canada and the United States.

* ***Victims of corporate downsizing***
  This would be similar to the '***New Entrants to the IT Workforce***'

Develop Your Personal People Network

Developing and maintaining an effective people network is vital to your career as an IT Contractor. As the tune made famous by the rock group the Beetles reminds us that "I get by with a little help from my friends." Personal relationships are vital for getting a second opinion, getting advice from a colleague with expertise in an area that you may lack, but need direction on. People networks are also vital in the search for the next contract assignment, in addition to your recruiter.

It has been my experience and that of dozens of my colleagues that most of our contracts are found through a friend who tells us of an upcoming contract opportunity in their company. You would still need to (in most cases) go through an agency, but being informed about opportunities, helps to keep you ahead of the game

The unwritten rule of the People Network is: You take from the Network, You give to the Network. Be prepared to return the favor when someone in your Network calls upon your expertise in a given area.

Chapter Review:

5-1)   Find (3) recruiting firms and find out their rate for: a) Java Programmer; b) Project Manager; c) Helpdesk Technician; d) Network Administrator; e) Network Engineer; f) System Analyst; g) Web Designer

5-2)   Search the internet for the average salary on the jobs in question (5-1).

Chapter Notes

6.0   Money Matters

Fact: *It's not how much money you make; it is how much you manage to keep in your coffers that count.*

Note: *All financial matters should be reviewed by your financial and/or tax professional to ensure that you are doing the best thing for you and taking advantage of all legal tax deductions that are applicable to your state or province.*

No matter how much you love what you do: money matters. You need it to pay for your living expenses, getting to and from work, vacation, and look after your family, and so on.

This chapter examines ways of keeping and optimizing your hard earned money.

It all boils down to discipline. IT is a hot and cold business. You can work contract after contract for two years, and then find yourself without a contract anywhere from one week to several months.

How much of your income that you keep, is much more important than how much you earn. Note, it is better to make $40,000 a year and be able to sock away $8,000 than to make $90,000 and have nothing to show for it. Forget about showing off about a high five or six figure income, you want (or should want) to be able to amass some wealth with the lowest tax impact as possible.

How? I will describe some simple, effective ways that anyone can take advantage of. Most of this info will apply to American and Canadian citizens; though most others can glean the basics and learn more about what is specific to their country of residence. It would be wise to consult an accountant for specific details that would apply to one's home province or state.

Retirement Saving Plans

In the US the 401K is the main savings plan used by W2 (salaried) type contractors. 1099 (independent / self-employed) contractors should consider using a traditional IRA's or Roth IRA's. These plans are designed to provide savings growth through tax-free interest earning (401K & traditional IRA) until it is withdrawn. For Canadian IT contractors the RRSP is the retirement savings plan that provides savings of pretax dollars for the purpose of retirement plus lower ones taxable income.

As much as possible, one should try to max out any & all tax advantages afforded them for a few reasons. First, it provides for savings for retirement years. Second, it provides emergency funds in case of hardship (longer than expect lag time between assignments, serious illness, etc.). In both the US & Canada at the time of writing, removal of retirement funds was treated as income and taxed as earnings to you within that tax year.

Savings Bonds & NET Banks

Both US & Canadian savings bond are a good investment for a few reasons. First, they're redeemable at any bank (if/when needed). Secondly, they are guaranteed by, you guessed it, the government. They are also willable to next of kin or (anyone you choose) in the event that you meet an untimely end. IT Contractors & net banks (internet banks) should be synonymous.

Chapter Review:

6-1)   Why should you have a retirement saving plan?

6-2)   How much of your income should you save?

6-3)   Who should you always pay first? Why?

Chapter Notes

'

7.0   Professionalism

What type of person do you consider a professional? We typically think of someone who is:

- Ethical
- Knowledgeable
- Curious
- Well dressed
- Skillful

Ethics is of vital importance to your continued success. As an IT contractor, your professionalism, in some cases more than your skill, will make or break your IT contracting career. In this book, ethics is the doing of what is 'morally' right. It is a unwritten code of conduct that you will become known by. Note too, that just because something is legal (or by the book), does not mean that it is ethical.

To be ethical and make yourself trustworthy by your peers, you must always strive to be beyond reproach. The items that will next be described are suggestions that have worked for many of my seasoned IT contractor's colleagues and myself. Keep in mind that you don't have to do it this way. These suggestions based on years of experience usually have a positive effect and helps to foster sound professional habits. Let's look at a few examples:

Pro Tip #1: Work Schedule

Arrive at work approximately 15-20 minutes prior to the start of your shift and leave 5-10 minutes past your shift. This gives the impression that you care about your assignment. By timing it so that you usually arrive early should cover you in the event that you cannot help being late one morning (traffic, sick child, etc.), it will be viewed as the exception instead of the rule.

Pro Tip #2: When Your Ideas are Rejected

You are normally hired for your expertise in a given area. If you are asked, assigned or encouraged to present your ideas, don't take it personally if your ideas are ever soundly rejected. Remember: 'The customer is always right!'. Insisting on getting it your way will not win you any points. It could also get you fired. There are times that you should/must stand your ground, but this is not one of them.

Pro Tip #3: Office Politics

You have no opinion. I repeat: *You have no opinion*. Do
not take any side. You have nothing to gain, except
resentment and a pink slip.

Pro Tip #4: Observing Office Culture

Observe and respect the way things are done in an office.
Even if they circulate hardcopy memos, when you know
that email is the more efficient way, let it go. As a
freelancer, you're there for a short time, not a long time –
get it?!

Pro Tip #5: CYA

Cover Your Assets; that means you. The IT field
unfortunately can be (make that '*is*') a cutthroat place.
Amongst professionals it should not be that way; reality
dictates otherwise. The rest of this chapter deals with a few
ways to make sure that you can always back yourself up
while doing your job.

KEEP DETAILED NOTES!!! This above all else will help reduce the amount of flack that may occur on the job. There are several ways to do this. A few effective ones are:

1)   E-mail storage;

2)   Keep an ASCII log file;

3)   Keep a work diary;

4)   Maintain a scrapbook;

5)   Desk calendar log;

6)   Commendation letter file.

There are other methods; however, amongst my colleagues and from personal experience, these methods have been found to be most effective. Whichever method you use be consistent. Let's examine each one's usefulness.

1) E-mail Storage

Keeping any important (translation: very controversial) e-mail correspondence. Many places use Lotus Notes or Microsoft Exchange for email services. I would strongly suggest keeping a printout of any emails that specifically talk about permission or the right to perform a certain action; getting those accountable to sign if necessary. Notices that I said 'keep a printout'. I know what you're thinking: 'Arg!! No, not paper!'; but it too is necessary at times,

2) Keep an ASCII Log File

Yes, for all those who are strict GUI interface people; think again. Many consultants use this CYA technique. Keeping an ASCII text log file of events that happen on a disk or pen drive makes sense for the following reasons: 1. It requires very little space; 2. Can be read by any word-processing program or text editor; 3. Can be reviewed/edited from a DOS prompt if need be.

3) Keep a Diary

Keep one in any form: ASCII text file (see previous), a diary book (i.e. 2003 desk diary), or a Lotus Notes or MS Outlook journal file. The main jest it to keep track of things so that you have a defense when cornered on thorny issues that you are sure you are innocent of. This is one case a PDA becomes a key piece of equipment. On a Palm or

Pocket PC you can keep a project diary and as an added
bonus, have it at you fingertips for instant recall.

4) Maintain a Scrapbook

A scrapbook or folder is a good idea if you are on a high
profile project that gets lots of media coverage. You can
keep newspaper and magazine clippings. If you are working
on a project team, gathering these things and organizing
them into a presentation can be used as self-promotion for
yourself on winning that next contract assignment. Using
media covered for this is best since you will not be
divulging any trade secrets; you are just using public
knowledge.

5) Desk Calendar Log

If you are fortunate enough to have some desk space on
your contract assignment, using a large calendar (desk
blotter or wall type) is useful for two reasons. You can keep
track of meetings, appointments and due dates at a glance
for the month. Secondly, by training those you work with to
check your desk calendar, everyone who needs to know
where you are (or forgets where you are) can use this to

answer many of their questions. For this to work effectively, you must keep it up-to-date.

6) Commendation Letter File

No one is going to (or should) promote you and your accomplishments better than you. If you get letters of thanks or commendation, keep them in a file folder that is readily accessible for interviews and or merit reviews.

Note: *Items 1-3 above can be kept on a Palm or Pocket PC*

Understanding the usefulness of each of these tools coupled with knowledge of your work environment will help you determine which tool is best for a given task. Over time you will develop your own personal favorites. I won't tell you mine less I jade your judgment. The tools are there, use the one(s) that work best for you in a given situation.

Chapter Review:

7-1)   Your agency assigns you as an IT contractor to 'Company X' to work as a business analyst on their data-mining project. You are doing things as you are instructed. You notice that the special program that is being written to access the data could be done quicker and for less cost using a customized web browser. Out of courtesy you present your idea. It is flatly rejected. Being the contracted 'expert', should you insist that it go your way?

7-2)   What program would you use for keeping an ASCII project log?

7-3)   What things should you not use email for on any clients sight?

7-4)   In what way(s) could you turn a project scrapbook into a presentation?

Chapter Notes

8.0   Vacation!!!

Remember the old saying 'All work and no play makes Jack a dull boy'? It is true, we all need a break now and then to clear the cobwebs, avoid going to work dressed in fatigues; you get the picture.

When I am not writing or contracting, I'm scuba diving (it's what I love to do). It would stand to reason then, that most of my favorite places on earth, Montego Bay (Jamaica), Cancun (Mexico), Miami (Florida, USA), are great scuba diving locations. I am always scheming on how to get back to one of these spots to chill and recharge my batteries (figuratively speaking). One favorite destination is not a dive spot – Montreal, Canada (got to have a little variety).

Whether you are the type who needs a few weeks or just a couple of days, you should work that time into you contract schedule. Since most contracts pay on an hourly basis, the end of a contract is usually a good point to spend the time with family or (for you single types) kick back and chill for a week or two.

While you are working on a contract with a typical duration (usually 3, 6 or 12 months), plan to be unemployed for a couple of months when your assignment ends. That means, you need to plan for being off financially speaking. This time off, welcomed or not should be a time to kick back, relax and enjoy the time; reflect on the past contract experience.

Use Long Weekends Effectively

In most cases, you are engaged to a client on an hourly bases and as such are not paid for any holidays that come up. You could (and should) plan for it so that you can put this time to good use. Here are a few suggestions:

1) Plan to be with family or friends

2) Do a non-IT related activity (i.e. zoo, art gallery, concert, etc.)

3) Read a good **non-IT** book

4) Go out of town (anywhere, even just to the next town) for a day or two

If you plan long weekends in this or a similar manor, you will end up going back to your assignment refreshed (if nothing else).

Chapter Review:

8-1) Why is vacation important?

8-2) How can you make good use of long weekends?

8-3) How can you make good use of time between assignments?

8-4) List 3 places that you would like to see and with whom. How do you see getting there?

Chapter Notes

About the Author

Courtney Thompson, is a freelance project manager, author, and lecturer. With a degree in Energy Conversion Technology, Mr. Thompson has over 20 years of technology experience with 16 of those years as an IT contractor. He encouraged many technology students to get a taste for contract work as the former Chairman of Computer Aided Design at a Chicago based Design College. He has written computer articles and has spoken at (2) international conferences. He also serves as an advisor on two technology boards for firms in Philadelphia, PA area.

Other Books by Courtney Thompson
'Palm @ Work & Play', ©2003
'Save Money in Any Economy', ©2003